~A BINGO BOOK~

Whole-Number Operations Bingo Book

COMPLETE BINGO GAME IN A BOOK

Place Value: Expanded Form

3,529,678

3 sets of one million	3,529,678
5 sets of one hundred thousand	3,529,678
2 sets of ten thousand	3,529,678
9 sets of one thousand	3,529,678
6 sets of one hundred	3,529,678
7 sets of ten	3,529,678
8 ones	3,529,678

Written By Rebecca Stark

Educational Books 'n' Bingo

TITLE: Whole-Number Operations Bingo
AUTHOR: Rebecca Stark

ISBN 978-0-87386-451-0

Educational Books 'n' Bingo

Printed in the U.S.A.

DIRECTIONS

INCLUDED:

List of Terms

Templates for Additional Terms and Clues

2 Clues per Term

30 Unique Bingo Sheets (To cut out or copy)

Sheet of Markers (to copy and distribute)

1. **Either cut apart the book or make copies of ALL the sheets. You might want to make an extra copy of the clue sheets to use for introduction and review. Keep the sheets in an envelope for easy reuse.**

2. Cut apart the call sheets with terms and clues.

3. Pass out one bingo sheet per student. There are enough unique sheets for a class of 30.

4. Pass out the markers. You may cut apart the markers included in this book or use any other small items of your choice. Students can also mark the sheets themselves; recopy the sheets as needed for additional games.

5. Decide whether or not you will require the entire sheet to be filled. Requiring the entire sheet to be filled provides a better review. However, if you have a short time to fill, you may prefer to have them do the just the border or some other format. Tell the class before you begin what is required.

6. There are 50 terms. Read the list before you begin. If there are any terms that have not been covered in class, you may want to read to the students the term and clues before you begin.

7. There is a blank space in the middle of each sheet. You can instruct the students to use it as a free space or you can write in answers to cover terms not included. Of course, in this case you would create your own clues. (Templates provided.)

8. Shuffle the sheets and place them in a pile. Two or three clues are provided for each term. If you plan to play the game with the same group more than once, you might want to choose a different clue for each game. If not, you may choose to use more than one clue.

9. Be sure to keep the sheets you have used for the present game in a separate pile. When a student calls, "Bingo," he or she will have to verify that the correct answers are on his or her sheet AND that the markers were placed in response to the proper questions. Pull out the sheets that are on the student's sheet keeping them in the order they were used in the game. Read each clue as it was given and ask the student to identify the correct answer from his or her sheet.

10. If the student has the correct answers on the sheet AND has shown that they were marked in response to the *correct questions,* then that student is the winner and the game is over. If the student does not have the correct answers on the sheet OR he or she marked the answers in response to *the wrong questions,* then the game continues until there is a proper winner.

11. If you want to play again, reshuffle the sheets and begin again.

Have fun

TERMS/ANSWERS

24	FACTORS
35	HUNDREDS PLACE
42	HUNDREDTHS PLACE
49	INTEGERS
54	INVERSE OPERATIONS
56	LONG DIVISION
63	MINUEND
72	MULTIPLICATION
81	ONE
100	ONES PLACE
1,000	PARENTHESES
ADDENDS	PLACE VALUE
ADDITION	PRODUCT
ASSOCIATIVE PROPERTY	QUOTIENT
BORROW	REMAINDER
CARRY	ROUND
COMMUTATIVE PROPERTY	SQUARE NUMBER
DECIMAL POINT	SUBTRACTION
DIFFERENCE	SUBTRAHEND
DIGIT	SUM
DISTRIBUTIVE PROPERTY	TENS PLACE
DIVIDEND	TENTHS PLACE
DIVISION	THOUSANDS PLACE
DIVISOR	WHOLE NUMBERS
ESTIMATE	ZERO

Clues for
Additional Terms

Write three clues for each of your additional terms.

_____	_____
1.	1.
2.	2.
3.	3.
_____	_____
1.	1.
2.	2.
3.	3.
_____	_____
1.	1.
2.	2.
3.	3.

Whole-Number Operations Bingo

Additional Terms

Choose as many additional terms as you would like and write them in the squares. Repeat each as desired.
Cut out the squares and randomly distribute them to the class.
Instruct the students to place their square on the center space of their card.

Whole-Number Operations Bingo

+ − X ÷	+ − X ÷	+ − X ÷	+ − X ÷	+ − X ÷
+ − X ÷	+ − X ÷	+ − X ÷	+ − X ÷	+ − X ÷
+ − X ÷	+ − X ÷	+ − X ÷	+ − X ÷	+ − X ÷
+ − X ÷	+ − X ÷	+ − X ÷	+ − X ÷	+ − X ÷
+ − X ÷	+ − X ÷	+ − X ÷	+ − X ÷	+ − X ÷
+ − X ÷	+ − X ÷	+ − X ÷	+ − X ÷	+ − X ÷
+ − X ÷	+ − X ÷	+ − X ÷	+ − X ÷	+ − X ÷

24 1. $(2 \times 4) \times 3 =$ 2. $(9 + 3) + 12 =$ 3. $(2 \times 6) + (3 \times 4) =$	35 1. $7 \times 5 =$ 2. $(5 \times 3) + (5 \times 4) =$ 3. $(7 \times 3) + 14 =$
42 1. $7 \times 6 =$ 2. $(3 + 3) \times (4 + 3) =$ 3. $100 - 58 =$	49 1. $7 \times 7 =$ 2. $(4 + 3) \times (5 + 2) =$ 3. $(10 \times 5) - 1 =$
54 1. $9 \times 6 =$ 2. $162 \div 3 =$ 3. $40 + 14 =$	56 1. $8 \times 7 =$ 2. $110 \div 2 =$ 3. $1\text{-}0 - 44 =$
63 1. $9 \times 7 =$ 2. $126 \div 2 =$ 3. $100 - 37 =$	72 1. $9 \times 8 =$ 2. $144 \div 2 =$ 3. $69 + 3 =$
81 1. $9 \times 9 =$ 2. $(12 \times 5) + (7 \times 3) =$ 3. $(3 \times 3) \times 9 =$	100 1. $100 \times 1 =$ 2. $1{,}000 \div 10 =$ 3. $10 \times 10 =$

Whole-Number Operations Bingo

1,000	**Addends**
1. 10,000 ÷ 10 =	1. The number being added in an addition problem.
2. 100 x 10 =	2. In the statement 5 + 2 = 7, the 5 and the 2 are these.
3. 100,000 ÷ 100 =	3. In the statement 100 + 300 = 400, the 100 and the 300 are these.

Addition	**Associative Property**
1. The operation in which two or more numbers are united.	1. Sometimes called the Grouping Property.
2. One of the four basic operations of mathematics. The other three are subtraction, multiplication and division.	2. This property states that changing the grouping of the addends does not change the sum: (3 + 2) + 4 = (4 + 3) + 2.
3. A plus sign is used to signify this operation.	3. It states that changing the grouping of the factors doesn't change the product: (5 x 2) x 3 = (2 x 3) x 5.

Borrow	**Carry**
1. In subtraction we often ___ 10 from the next higher digit column to obtain a positive difference in the column in question.	1. To transfer a number from one column of figures to the next.
2. To solve 20 – 19 ___ 10 from the tens column to get a positive difference in the ones column.	2. To solve 12 x 5, you must ___ one set of ten to the tens column.
2. To solve 234 – 143, ___ from the 100s column to get a positive difference in the tens column.	2. To solve 16 x 4, you must ___ two sets of ten to the tens column.

Commutative Property	**Decimal Point**
1. Sometimes called the Order Property.	1. A symbol used to separate the ones place from the tenths place.
2. It states that changing the order of the addends doesn't change the sum: 2 + 3 = 3 + 2.	2. In money it separates dollars from cents.
3. It states that changing the order of the factors doesn't change the sum: 2 x 3 = 3 x 2.	3. To multiply a number by 10, move this one place to the right.

Difference	**Digit**
1. The result when one number is subtracted from another.	1. Any of the numerals 1 to 9 and 0.
2. In the statement 10 – 4 = 6, 6 is the ___.	2. 36 is a 2-___ number.
3. In the statement 500 – 90 = 410, 410 is this.	3. 268 is a 3-___ number.

Whole-Number Operations Bingo

© Barbara M. Peller

DISTRIBUTIVE PROPERTY
1. This property lets you multiply a sum by multiplying each addend separately and then adding the products.
2. To use the ___ to multiply 3 x 42, multiply (3 x 40) and (3 x 2) and add the results.
3. An example of the ___ is
5 x (3 + 4) = (5 x 3) + (5 x 4).

Dividend
1. The number to be divided in a division problem.
2. In the statement 60 ÷ 5 = 12, 60 is this.
3. In the statement 72 ÷ 9 = 8, 72 is this.

Division
1. The process of determining how many times one number is contained in another.
2. One of the basic operations. The other 3 are addition, subtraction and multiplication.
3. A line with a dot above and below it (called an obelus or ___ sign) is sometimes used. In long ___ a bracket is used.

Divisor
1. The number by which the dividend in a division problem is divided.
2. In the statement 60 ÷ 5 = 12, 5 is this.
3. In the statement 72 ÷ 9 = 8, 9 is this.

Estimate
1. As a verb, it means "to guess."
2. An approximate calculation of quantity, size, degree or value.
3. To determine roughly the size, quantity or value.

Factors
1. Numbers multiplied together.
2. In the statement 4 x 7 = 28, 4 and 7 are these.
3. In the statement 9 x 6 = 54, 9 and 6 are these.

Hundreds Place
1. The place three to the left of the decimal point.
2. In the number 247, the 2 is in this place.
3. In the number 5,367, the 3 is in this place.

Hundredths Place
1. The place to the right of the tenths place.
2. The number two to the right of the decimal point.
3. In the number 25.692, the 9 is in this place.

Integers
1. The positive whole numbers, the negative whole numbers and zero.
2. These positive and negative numbers and zero can be shown on a number line.
3. The terms positive and negative can only apply to these.

Inverse Operations
1. Operations that undo the effect of each other.
2. Addition and subtraction have this kind of relationship.
3. Multiplication and division have this kind of relationship.

Long Division 1. Division that requires each step to be written out. 2. When performing this operation we use a bracket instead of an obelus. 3. A multi-step process needed when the divisor is a large number.	**Minuend** 1. The number from which another number is subtracted. 2. In the statement 10 – 4 = 6, 10 is this. 3. In the statement 500 – 90 = 410, 500 is this.
Multiplication 1. This operation is really repeated addition. 2. An "x," or times sign, is used. 3. We use this process to solve the problem 6 x 5 = ?	**One** 1. Any number times this is that number. 2. 54,798 x this number is 54,798. 3. The lowest possible positive odd number.
Ones Place 1. The place just to the left of the decimal point. 2. In the number 247, the 7 is in this place. 3. In the number 5,367, the 7 is in this place.	**Parentheses** 1. Indicates terms that are one unit. 2. Symbols used in pairs to group things together. 3. First perform any calculations inside these symbols first.
Place Value 1. Value given to a particular figure because of its place in the number. 2. In the number 624, the __ of 6 is "hundreds." 3. In the number 624, the __ of 2 is "tens."	**Product** 1. The answer to a multiplication problem. 2. The result when two or more factors are multiplied. 3. In the statement 7 x 4 = 28, 28 is this.
Quotient 1. The answer to a division problem. 2. In the statement 60 ÷ 5 = 12, 12 is the ___. 3. In the statement 72 ÷ 9 = 8, 8 is the ___.	**Remainder** 1. If a division problem does not work out evenly, the number left over is this. 2. In the problem 12 ÷ 5, the __ will be 2. 3. In the problem 73 ÷ 9, the __ will be 1.

Whole-Number Operations Bingo

Round	**Square Number**
1. We do this to eliminate one or more digits farthest to the right when precision is not required. 2. To __ to a whole number, you would change 49.658 to 50. 3. To __ to the nearest tenth, you would change 49.658 to 49.7.	1. Result when both factors in a multiplication sentence are the same. 2. The product of 5 x 5 is a __. 3. Because both factors are the same, the product of 130 x 130 will be a __.
Subtraction	**Subtrahend**
1. The process of finding the difference between two numbers. 2. The opposite operation of addition. 3. A minus sign is used to signify this operation.	1. The number being subtracted from another. 2. In the statement 10 – 4 = 6, 4 is the ___. 3. In the statement 500 – 90 = 410, 90 is this.
Sum	**Tens Place**
1. The result of adding numbers together. 2. In the statement 5 + 2 = 7, the 7 is this. 3. In the statement 100 + 300 = 400, the 400 is this.	1. The place two to the left of the decimal point. 2. In the number 247, the 4 is in this place. 3. In the number 5,367, the 6 is in this place.
Tenths Place	**Thousands Place**
1. The place just to the right of the decimal point. 2. In the number 16.275, the 2 is in this place. 3. In the number 25.692, the 6 is in this place.	1. The place just to the left of the hundreds place. 2. In the number 8,627, the 8 is in this place. 3. In the number 25,692, the 5 is in this place.
Whole Numbers	**Zero**
1. Positive numbers without a fraction or a decimal point. 2. 5, 10, 312 and 447 are __. -2, 10.5 and 3/4 are not. 3. 7; 12; 1,312; and 24,447 are __. -284, 14.5 and 0 are not.	1. It means "none" and is neither positive nor negative. 2. The sum of any number plus this is that number. 3. The product of any number times this is zero.

Whole-Number Operations Bingo

Long Division	24	54	Estimate	Distributive Property
Commutative Property	35	Zero	Addends	Hundredths Place
49	Thousands Place		Product	Inverse Operations
56	Carry	Tenths Place	Multiplication	Place Value
Quotient	Dividend	Decimal Point	Square Number	Hundreds Place

Whole-Number Operations Bingo

56	49	Minuend	Sum	Addition
Place Value	Addends	72	Carry	One
100	Dividend		Difference	Tenths Place
Round	Subtrahend	Thousands Place	Subtraction	Hundreds Place
Hundredths Place	Zero	Decimal Point	Commutative Property	Square Number

Whole-Number Operations Bingo

56	Tenths Place	Addends	Multiplication	49
Dividend	35	81	24	Factors
Carry	Zero		One	42
Thousands Place	100	Quotient	Round	Minuend
Square Number	Commutative Property	Decimal Point	Subtraction	Addition

Whole-Number Operations Bingo

Thousands Place	One	54	Commutative Property	Addition
Integers	63	24	Sum	49
Product	Round		Distributive Property	Estimate
Tenths Place	Digit	Zero	Decimal Point	72
Associative Property	Hundredths Place	Whole Numbers	Square Number	Inverse Operations

Whole-Number Operations Bingo

Hundredths Place	Distributive Property	Carry	72	Commutative Property
Integers	Tenths Place	81	Difference	35
54	Inverse Operations		Parentheses	Division
Hundreds Place	Addition	Long Division	Subtraction	1,000
Addends	Decimal Point	49	Thousands Place	Product

Whole-Number Operations Bingo

42	One	Minuend	Addition	Inverse Operations
Multiplication	Carry	1,000	24	49
Sum	Associative Property		63	Difference
Decimal Point	Quotient	Subtraction	Whole Numbers	54
Place Value	Tenths Place	Long Division	Product	Digit

Whole-Number Operations Bingo

Long Division	One	Division	Parentheses	Addends
Place Value	Addition	Dividend	35	Integers
Minuend	Estimate		Difference	63
Thousands Place	Round	81	56	100
Decimal Point	Commutative Property	Subtraction	Whole Numbers	42

Whole-Number Operations Bingo

Product	One	Borrow	Multiplication	63
Integers	54	Sum	Inverse Operations	72
Digit	Remainder		Addition	Distributive Property
Square Number	Thousands Place	56	Associative Property	Round
Zero	Decimal Point	Whole Numbers	Carry	Place Value

Whole-Number Operations Bingo

Difference	Addends	Dividend	Digit	Commutative Property
Associative Property	Addition	Product	Carry	One
Factors	Long Division		35	Borrow
1,000	Hundreds Place	Quotient	Parentheses	Division
Round	Subtraction	81	56	Distributive Property

Whole-Number Operations Bingo

56	Multiplication	63	Sum	Digit
Inverse Operations	72	24	35	Addition
Remainder	One		Estimate	100
Quotient	Hundreds Place	1,000	Subtraction	Factors
81	Place Value	Minuend	Hundredths Place	Product

Whole-Number Operations Bingo

42	One	Carry	1,000	Place Value
Borrow	Factors	Parentheses	Difference	24
Integers	Addition		Minuend	Dividend
81	49	Subtraction	Commutative Property	56
Associative Property	Decimal Point	Long Division	Whole Numbers	Addends

Whole-Number Operations Bingo

Addends	Distributive Property	Factors	Multiplication	Difference
Dividend	Place Value	54	Whole Numbers	35
Long Division	Division		Inverse Operations	Sum
Decimal Point	Round	Addition	56	Integers
One	Borrow	Remainder	Associative Property	72

Whole-Number Operations Bingo

1,000	Distributive Property	42	Factors	Inverse Operations
54	Borrow	Addition	Difference	100
Multiplication	72		Dividend	Division
Product	Subtraction	63	Remainder	56
Decimal Point	Hundreds Place	Whole Numbers	Long Division	Parentheses

Whole-Number Operations Bingo

Commutative Property	Addition	Carry	Difference	Associative Property
72	Long Division	Factors	35	One
1,000	Estimate		Minuend	81
Hundreds Place	Subtraction	Remainder	63	42
Decimal Point	Sum	100	Place Value	Product

Whole-Number Operations Bingo

Parentheses	Difference	Carry	Addends	Multiplication
42	Minuend	24	54	Associative Property
Inverse Operations	Long Division		49	One
Decimal Point	Factors	Borrow	Subtraction	1,000
Place Value	Round	Whole Numbers	Digit	Dividend

Whole-Number Operations Bingo

63	Factors	Borrow	Digit	Subtrahend
Sum	100	Division	Integers	Estimate
1,000	Distributive Property		Inverse Operations	Dividend
Thousands Place	72	Decimal Point	Parentheses	56
Associative Property	Tens Place	Whole Numbers	Round	One

Whole-Number Operations Bingo

81	Ones Place	Divisor	Factors	Commutative Property
Parentheses	Associative Property	Subtraction	Estimate	Division
Difference	Product		Tens Place	Borrow
Hundreds Place	Place Value	56	Carry	100
Quotient	1,000	Addends	Multiplication	Distributive Property

Whole-Number Operations Bingo

Digit	Remainder	72	1,000	Sum
One	81	Quotient	Inverse Operations	Associative Property
Difference	100		Divisor	54
Hundreds Place	24	Subtraction	56	Minuend
Tens Place	Factors	Carry	Ones Place	42

Whole-Number Operations Bingo

Inverse Operations	42	Factors	Borrow	Remainder
Parentheses	Multiplication	One	Addends	Estimate
Ones Place	Commutative Property		35	49
Minuend	Tens Place	Quotient	Round	Divisor
54	Subtrahend	Place Value	Product	Whole Numbers

Whole-Number Operations Bingo

Remainder	Ones Place	Multiplication	Factors	Whole Numbers
72	Dividend	Integers	Quotient	Sum
Distributive Property	Division		Thousands Place	24
Hundredths Place	Zero	Square Number	Round	Tens Place
Tenths Place	Product	Subtrahend	56	Divisor

Whole-Number Operations Bingo

Parentheses	42	Integers	Factors	Hundredths Place
Distributive Property	Divisor	63	Borrow	Long Division
100	Place Value		Ones Place	Carry
Quotient	Addends	Tens Place	Hundreds Place	Product
Thousands Place	Subtrahend	Whole Numbers	81	Round

Whole-Number Operations Bingo

Digit	Minuend	Divisor	54	1,000
Sum	Multiplication	49	Borrow	35
72	Estimate		Long Division	Division
Tens Place	Hundreds Place	Round	24	Commutative Property
Subtrahend	81	Ones Place	100	Integers

Whole-Number Operations Bingo

63	Ones Place	Addends	54	Whole Numbers
42	Remainder	Place Value	Parentheses	24
Minuend	1,000		Square Number	Long Division
100	Subtrahend	Tens Place	81	Round
Hundredths Place	Zero	Product	Quotient	Divisor

Whole-Number Operations Bingo

63	Remainder	Commutative Property	Ones Place	Borrow
Divisor	Whole Numbers	Integers	Sum	Long Division
Division	Digit		1,000	100
Hundredths Place	Square Number	Tens Place	81	Distributive Property
Tenths Place	Thousands Place	Subtrahend	Multiplication	Zero

Whole-Number Operations Bingo

Thousands Place	Integers	Ones Place	Carry	Divisor
24	Hundreds Place	Parentheses	63	35
Distributive Property	Borrow		Square Number	Tens Place
49	Hundredths Place	Zero	Subtrahend	Estimate
Whole Numbers	Commutative Property	72	Associative Property	Tenths Place

Whole-Number Operations Bingo

Divisor	Ones Place	Minuend	Sum	Digit
Quotient	Multiplication	Borrow	Remainder	63
Hundreds Place	Square Number		Estimate	Thousands Place
81	54	Hundredths Place	Subtrahend	Tens Place
Division	Associative Property	Carry	Zero	Tenths Place

Whole-Number Operations Bingo

Minuend	72	Ones Place	Remainder	Dividend
Hundredths Place	Square Number	Parentheses	Tens Place	35
Subtraction	Zero		Subtrahend	Thousands Place
Digit	42	Integers	Tenths Place	24
Associative Property	Estimate	Divisor	49	Division

Whole-Number Operations Bingo

Inverse Operations	Remainder	49	Ones Place	63
Dividend	Divisor	Square Number	Sum	Estimate
Zero	100		Division	Quotient
56	Digit	Place Value	Subtrahend	Tens Place
54	Difference	Associative Property	Tenths Place	Hundredths Place

© Barbara M. Peller

Whole-Number Operations Bingo

Divisor	Remainder	Digit	Parentheses	Difference
Hundreds Place	Quotient	Integers	Division	49
Distributive Property	Square Number		35	Ones Place
Dividend	Hundredths Place	Addition	Subtrahend	Tens Place
63	Borrow	Tenths Place	42	Zero

Whole-Number Operations Bingo

Commutative Property	Ones Place	Sum	Difference	Tens Place
24	Remainder	Minuend	Estimate	35
Hundreds Place	1,000		Division	Integers
Tenths Place	42	54	Subtrahend	Square Number
Hundredths Place	Addends	Zero	Divisor	49

www.ingramcontent.com/pod-product-compliance
Lightning Source LLC
Chambersburg PA
CBHW051428200326
41520CB00023B/7402